CIA

Field Expedient Methods for Explosives Preparation

Central Intelligence Agency

TABLE OF CONTENTS

FIELD EXPEDIENT METHODS FOR EXPLOSIVES PREPARATION

PREPARATION OF LEAD PICRATE

Lead Picrate is used as a primary explosive in the fabrication of detonators. It is to be used with a booster explosive such as picric acid or RDX.

MATERIAL REQUIRED (For 1 detonator):

Litharge (lead monoxide) (field prepared; also is used in plumbing and ceramic cements)

Picric acid (field prepared)

Wood alcohol (methanol) (some antifreezes and paint removers)

Wooden or plastic rod

Dish or saucer (china or glass)

Teaspoon

Improvised scale (see Section VII, No. 8)

PROCEDURE:

1. Weigh out 2 grams (or equal amounts) by weight of picric acid and lead monoxide.

2. Add the picric acid to 2 teaspoons (10 milliliters) of methanol in a container (dish or saucer) and stir.

3.   Add lead monoxide and stir.

4. Continue stirring and allow the alcohol to evaporate. NOTE: The mixture will suddenly thicken,

5. Carefully break up this mixture and stir occasionally until a powder is formed (a few lumps will remain).

6. Remove and spread out to air dry. NOTE: if possible, dry at 100°C (212 F) for two hours.

PREPARATION OF PICRIC ACID
FROM ASPIRIN

Picric acid can be used as a booster explosive in detonators or as an intermediate to preparing lead picrate or DDNP.

MATERIAL REQUIRED:

Aspirin, 20 tablets (5 grain/tablet)

Alcohol - 95 percent pure

Concentrated sulfuric acid (boil battery acid until white fumes appear)

Potassium nitrate (saltpeter)

Water

Canning jar (1 pint)

Hot water bath

Paper towels (for filter)

Glass tube or rod

Glass containers

Dish (ceramic or glass)

Heat source

Cup

Teaspoon

PROCEDURE:

1. Crush 20 tablets of aspirin in a glass container and work into a paste with a teaspoon of water.

2. Add approximately 1/3 - 1/2 cup (100 milliliters) of alcohol with stirring and filter through a paper towel into another glass container.

   If aspirin is pure enough (usually cheap priced aspirin are) then Steps 1, 2 and 3 can be omitted and can proceed and add crushed aspirin to Sulfuric Acid.

3. Discard the solid left on the paper and pour the liquid from the container into the dish. Evaporate the alcohol and water on a hot water bath, leaving a white powder.

4. Add this white powder to 1/3 cup (80 milliliters) of concentrated sulfuric acid in a canning jar.

5. Heat the jar in a simmering hot water bath for 15 minutes and remove. Stir; solution will gradually turn black.

6. Add 3 level teaspoons (15 grams) of potassium nitrate in three portions with vigorous stirring. After heating put jar in cold H_2O and add KNO_3.

7. Allow the yellow-orange solution to cool to room temperature with intermittent stirring. Solution is darker than yellow-orange.

8. Pour the solution into 1-1/4 cups (300 milliliters) of water (cold if possible) and allow to cool. Upon pouring, solution is now yellow-orange.

9. Filter the mixture through a paper towel and wash the light yellow material (picric acid) with 1/8 cup (25 milliliters) of water through paper towel. Takes considerable time to filter if genuine filter paper is used.

10. Dry at 160-200 °F for two hours. The yield is 4.0 - 4.5 grams.

TETRAMMINECOPPER (II) CHLORATE

Tetramminecopper (II) chlorate is a primary explosive that can be made from sodium chlorate, copper sulfate, and ammonia. This explosive when used with a booster is capable of detonating composition C-4.

MATERIAL REQUIRED:

Sodium chlorate (obtain as medicine, weed killer, defoliant, or by field method)

Copper sulfate (obtain as insecticide, water purifying agent, or by field method)

Ammonium hydroxide (obtain as household ammonia or smelling salts)

Alcohol - 95% pure

Wax or clay

Water

Narrow mouth bottle (wine or coke)

Wide mouth bottles (mason jars)

Tubing (rubber, copper, steel)

Teaspoon

Improvised scale (see Section VII , No. 8)

Heat source

Paper towel (for filter)

Pan

PROCEDURE:

1. Measure 1/3 teaspoon (2.5 grams) of sodium chlorate in a wide mouth bottle, then add 10 teaspoons of alcohol.

2. Add 1 teaspoon (4 grams) of copper sulfate and stir the mixture just under the boiling point for 30 minutes (heat can be supplied by a pan of hot water). The mixture will change color. NOTE: Keep solution away from flame.

3. Keep volume of the solution constant by adding additional alcohol about every 10 minutes. Remove solution and let cool. Filter through folded paper towels into another wide mouth bottle. Keep the liquid.

4. Add 1 cup (250 milliliters) of ammonia to the narrow mouth bottle. Placing tubing so that it extends about 1-1/2 inches inside bottle, then seal tubing to bottle with wax (pitch, clay, etc.).

5. Place tubing into solution from step (3) as indicated in the figure (next page). Heat bottle containing ammonia in a pan of hot water (not boiling) for about 10 minutes.

6. Bubble ammonia gas through solution until the color of the solution changes from a light green to a dark blue (approximately 10 minutes) and continue bubbling for another ten minutes.

▷ CAUTION: At this point the solution is a primary explosive. Keep away from flame.

7. Reduce the volume of the solution to about 1/3 of its original volume by evaporating in the air or in a stream of air.

8. Filter crystals through a folded paper towel and wash once with 1 teaspoon alcohol, dry overnight (16 hours).

▷ CAUTION: Explosive is shock and flame sensitive. Store in a capped container.

POTASSIUM (OR SODIUM) NITRITE AND LITHARGE (LEAD MONOXIDE)

Either of the nitrites are needed to prepare DDNP and litharge is required for the preparation of lead picrate.

MATERIAL REQUIRED:

Lead metal (small pieces or chips)

Potassium (or sodium) nitrate

Wood (methyl) alcohol

Iron pipe with closed end or cap (one end only)

Iron rod (screwdriver)

Paper towels

Glass jars, wide mouth (2 each)

Pan (metal)

Heat source (such as hot coals or blow torch)

Improvised scale (see Section VII, No. 8)

Cup

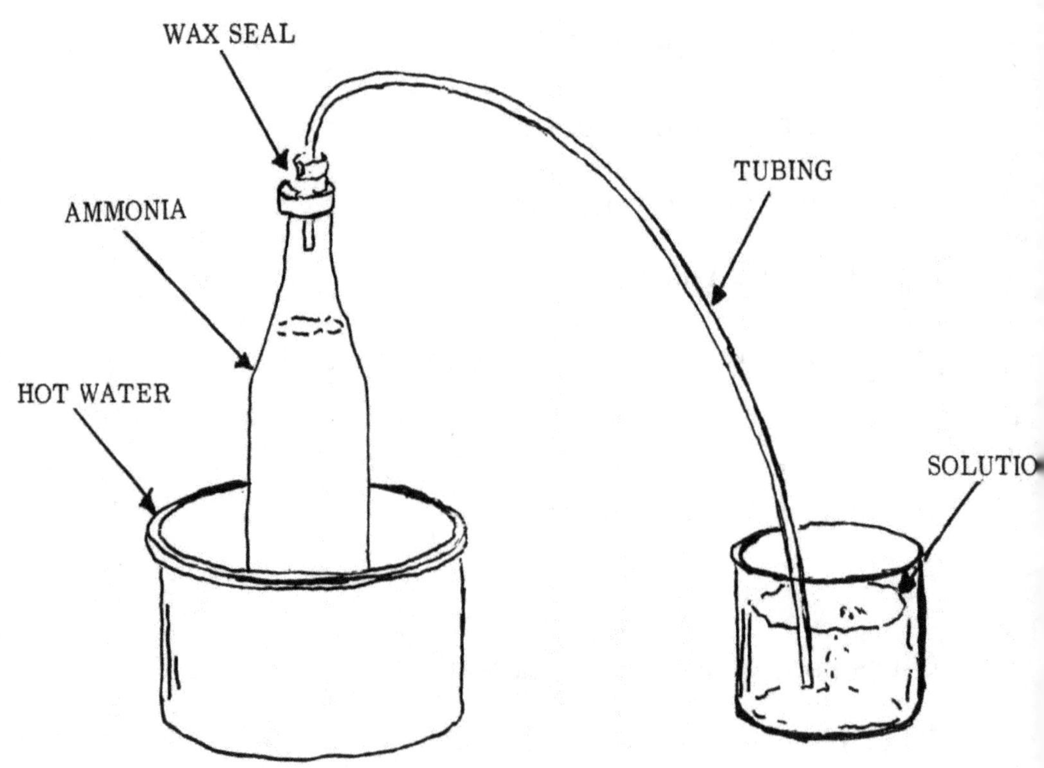

WAX SEAL
AMMONIA
HOT WATER
TUBING
SOLUTIO

PROCEDURE:

1. Mix 12 grams of lead and 4 grams of potassium (or sodium) nitrate. Place the mixture in the iron container (iron pipe) and heat in a hot bed of coals or by a blow torch for hour or more.

2. Remove the container and allow to cool. Chip out the yellow solid with the screwdriver and add to 1/2 cup (120 milliliters) of methyl alcohol in the jar. Orange-Brown solid, cream color in alcohol.

3. Heat the mixture in a pan of hot water until it reacts. Solution turns darker upon heating.

4. Filter the mixture through a paper towel into the second jar.

5. The solid left on the paper is lead monoxide. Wash it through paper twice, using 1/2 cup (120 milliliters) hot water each time and air dry before using in explosive preparation (for example, lead picrate).

6. Place the jar with the liquid in the hot water and heat until alcohol has evaporated. The remaining powder is the nitrite, snowy liquid; some white powder appears but not much.

NOTE: Sodium nitrite has a strong tendency to absorb water from the atmosphere and should be stored in a closed container.

PREPARATION OF COPPER SULFATE (PENTAHYDRATE)

Copper sulfate is a required material for the preparation of TAC.

MATERIAL REQUIRED:

Copper wire or pieces

Sulfuric acid (battery)

Potassium nitrate or nitric acid (field grades)

Alcohol

Water

Two heat resistant glasses or jars, one pint

Paper towels

Metal pan (for hot water bath)

Wooden rod

Improvised scale (see Section VII, No. 8)

Cup

PROCEDURE:

1. Place 10 grams of copper into one of the jars and add one cup (240 milliliters) of sulfuric (battery) acid. To this mixture add 12 grams of potassium nitrate, or 1-1/2 teaspoons of nitric acid. NOTE: Nitric acid gives a product of greater purity.

2. Heat the mixture on a hot water bath (near boiling) until the bubbling has ceased (requires about two hours).

CAUTION: This reaction evolves strong toxic fumes and therefore must be performed in an open, well ventilated area.

3. Pour the hot blue liquid into a second vessel (keeping the unreacted copper in the first jar) and allow to cool at room temperature.

4. After the crystals have formed, carefully pour away the liquid and break up the crystals. Then add 1/2 cup (120 milliliters) of alcohol to the powder and stir.

5. Pour the solution through the towel filter and wash the solid left on the paper three times, using 1/2 cup (120 milliliter) portions of the alcohol each time.

6. Allow the product to air dry for two hours.

RECLAMATION OF RDX FROM C-4

RDX can be used as a booster explosive for detonators or as a high explosive charge.

MATERIAL REQUIRED:

Gasoline

C-4

Wide mouth jars, one pint (2 each)

Paper towels

Wooden stirring rod

Teaspoon

PROCEDURE:

1.   Place 1 teaspoon (10 grams) of C-4 in the pint jar and add one cup (240 milliliters) of gasoline.

2. Knead and stir the C-4 with the rod until the C-4 has broken up. Allow to stand 1/2 hour.

3. Start the stirring again until a fine white powder remains on the bottom of the container.

4. Filter the mixture through a paper towel and wash the solid left on the paper with 1/2 cup (120 milliliters) of gasoline.

5. Air dry for several hours or heat at 80-100° C (about 150-212 °F) for one hour.

This Page Intentionally Left Blank

LABORATORY METHODS FOR PREPARING PROMISING EXPLOSIVES

LEAD AZIDE

Preparation of Hydrazinium Sulfate.

$$2 \ NH_3 + NaOCl + H_2SO_4 \longrightarrow N_2H_4 \cdot H_2SO_4 + NaCl + H_2O$$

One-hundred-forty-one ml of Chlorox bleach (5.25 precent NaOCl) was added to 200 ml of 20 percent ammonium hydroxide and 5 ml of 1 per cent limewater $Ca\,(OH)_2$ in a one liter Erlenmeyer flask. The mixture was rapidly heated to boiling and maintained until the volume was reduced to about half, which required about one-half hour. The solution was rapidly cooled and dilute sulfuric acid was added until a pH of 7-8 was attained and the precipitate that formed was separated by filtration. The cold filtrate was strongly acidified with 40 percent suffuric acid. The white precipitate was filtered, washed with methanol and air dried. Melting point 254 $\pm 2\,°C$ (lit. 254 °C).

Preparation of Isopropyl Nitrite.

A mixture of 45 ml concentrated sulfuric acid, 30 ml water and 110ml isopropyl alcohol, previously cooled to O°C, was added to an ice cold solution of 114 grains of sodium nitrite in 450 ml of H_2O. Slow addition required about two hours in order to maintain temperature around O°C. The upper oily layer was separated and washed three times with 30 ml portions of 5 gram 100 ml sodium bicarbonate solution and 22 grams NaCl 100 ml solution respectively.

Preparation of Sodium Azide.

$$N_2H_4 \cdot H_2O + NaOH + C_4H_9ONO_2$$
$$\longrightarrow NaN_3 + C_4H_9OH + 3H_2O$$

Five grams of caustic soda (NaOH) was dissolved in 50 ml of ethyl alcohol (3A), and the clear portion was decanted in a 100 ml distilling flask containing 6 ml of hydrazine hydrate. After adding one ml of butyl nitrite (or isopropyl nitrite) the mixture was heated on a steam bath to initiate the reaction. Twelve ml more of the nitrite was slowly added in such a manner that the mixture refluxed slowly. Addition required about one hour and the mixture was heated an additional fifteen minutes. The reaction flask was cooled and the solid product collected on a filter. The product was washed with alcohol and air dried. Recrystallization from water yielded white crystalline material.

Lead Azide.

The following solutions were prepared:

Solution A: 0.20 g of sodium azide and 0.006 g of sodium hydoxide in 7 ml of water.

Solution B: 0.69 grams $Pb(NO_3)_2$ and 0.04 grams Dextrin dissolved in a solution of 9 ml water. This solution is brought to a pH of 5 by the addition of dilute NaOH.

Solution B was brought to 60°C and Solution A was slowly added with stirring. The mixture was allowed to stir till ambient temperature was attained and the solid azide collected on a filter. After washing with water and air drying the product weighed 0.4 grams. This product was found capable of initiating RDX when incorporated into a No. 6 blasting cap.

NITROUREA

Urea Nitrate.

Twenty-five ml of concentrated nitric acid was added dropwise to an ice cold solution of 5.0 grams urea in 25 ml of water. The white precipitate was collected in a Buchner funnel and washed with ice cold nitric acid. After drying at 100°C for one hour, yield was 9.2 grams.

Nitrourea Preparation:

Nine (9.0) grams of the above product, urea nitrate, was added portionwise to 32 ml of ice cold (-3°C) concentrated sulfuric acid at such a rate that the temperature did not exceed 5°C. Total time for addition was approximately one-half hour, after which the mixture was poured on 75 grams of ice. The white precipitate was filtered, washed with ice cold water, just sufficient to cover it, and air dried. Material obtained weighed 5.3 grams and melted with decomposition at 157°-158°C (lit. 158-159°C).

MANNTTOL HEXANITRATE/ NITROMANNITOL

Ten grams of mannitol was added portionwise to 50 ml of concentrated nitric acid keeping the temperature at 0°C. One-hundred ml of concentrated sulfuric acid was then added dropwise maintaining temperature around 0°C. The slurry was filtered*, washed with water and a dilute $Na2CO_2$ solution and water again, until washings were neutral. Product was air dried and had a melting point of 113-115° C (lit. 112-113 °C).

LEAD NITRANILATE

Chloranil:

A slurry of 5.0 grams of salicylic acid and 100 ml of concentrated hydrochloric acid was heated to 80°C and 5.0 grams of potassium chlorate added in portions (with effervescence). An additional 400 ml of concentrated hydrochloric acid and 5.0 grams potassium chlorate was added and the mixture allowed to heat at 80-90°C for four hours. After filtering, washing with water and air drying, the yellow crystals melted at 190-200°C (sealed tube). The yield was 5.45 grams.

* In field use, the insoluble slurry which floats to the top of the solution, may be decanted or scooped off.

Sodium Nitranilate.

A mixture of 5.0 grams of chloranil and 200 ml of ethyl alcohol were heated to boiling and treated with a solution of 5.6 grains of sodium nitrite in 100ml of ethyl alcohol. The mixture was heated with stirring for one-half hour and allowed to cool. The orange-gold crystalline product was collected on a filter, washed with ethyl alcohol and air dried. Yield was 1.85 grams.

Lead Nitranilate.

A solution of 1.0 grams of sodium nitranilate in 100 ml of boiling was filtered and the filtrate treated with 2.9 grams of lead nitrate in 10 ml of water. The mixture was stirred for one-half hour and the gold platelets collected on a filter and with water. After drying at 80°C for three hours, the product weighed 1.23 grams. The product ignited with a loud report on flame contact.

CHLORATOTRIMERCURIACETALDEHYDE

A stream of acetylene (from calcium carbide) was led into a solution of 0.3 gram of mercuric nitrate and 0.1 gram sodium chlorate in 20 ml of water. The solid suspension was initially white then turned grey upon further treatment with acetylene. The product was collected on a filter, washed with water and air dried. Ignition temperature was ca. 150°C as determined with a Fisher-Johns melting point apparatus.

TRISHYDRAZINEZINC (II) NITRATE

To a solution of 5.0 grams of zinc nitrate in 25 ml of ethyl alcohol was added a solution of 1.7 ml of 95 percent hydrazine in 5 ml of alcohol. The precipitate that immediately formed was collected on a filter and washed with ethyl alcohol. After drying for 2.5 hours at 90°C the product weighed 4.45 grams.

FULMINATING SILVER

Silver chloride (from silver nitrate and hydrochloric acid) was treated with sufficient ammonium hydroxide to effect solution. This solution was treated with potassium hydroxide pellets until effervescence subsided. The black mixture was diluted with water and the dark solid collected on a filter. Attempts to remove the product from the wet filter resulted in an explosion.

DIAMMINESILVER (I) CHLORATE

A near saturated solution of silver chlorate (from silver nitrate and sodium chlorate) in water was treated dropwise with concentrated ammonium hydroxide. The dark product mildly deflagrated after isolation by filtration.

TRIACETONETRIPEROXIDE

A mixture of 5 ml of acetone and 10 ml of 6 percent hydrogen peroxide was cooled to 5°C and treated dropwise with 3 ml of concentrated sulfuric acid. The white precipitate that formed was extracted with ether and the extract was washed three times with cold water. The ether was evaporated by a stream of air and the product melted at 90-95° C. (lit. 94-95°C.)

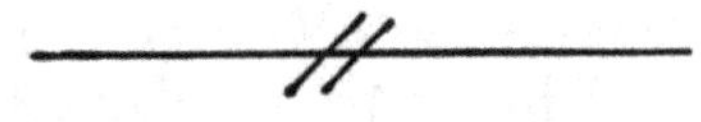

This Page Intentionally Left Blank

www.ingramcontent.com/pod-product-compliance
Lightning Source LLC
LaVergne TN
LVHW052208200726
843508LV00015B/1981